QUEEN VICTORIA

A PHOTOGRAPHIC JOURNEY

QUEEN VICTORIA

A PHOTOGRAPHIC JOURNEY

CHRIS FRAME AND RACHELLE CROSS

FOREWORD BY ALASTAIR GREENER

WITH AN AFTERWORD FROM
CAPTAIN CHRISTOPHER RYND

The
History
Press

For Vicki, Jan and John

First published 2010

The History Press

The Mill

Brimscombe Port

Stroud

Gloucestershire

GL5 2QG

www.thehistorypress.co.uk

British Library Cataloguing in Publication Data.
A catalogue record for this book is available from the British Library.

ISBN 978-0-7524-5298-2
Typesetting and origination by The History Press Ltd
Printed and bound in India by Replika Press Pvt. Ltd.
Manufacturing managed by Jellyfish Print Solutions Ltd

CONTENTS

FOREWORD

BY ALASTAIR GREENER

When Chris and Rachelle first mentioned the idea of a foreword for this book, I was honoured to be asked, as their previous books have done such a great job in capturing the essence of what the Cunard Queens are, and I was confident this would be a great tribute to our latest Cunarder. After eleven years at sea, I joined Cunard in 2006, and immediately loved the tradition and heritage associated with a company that began plying the oceans in 1840. To be associated with that history on the two most famous liners in the world – *Queen Mary 2* and *Queen Elizabeth 2* – was incredible. To be invited to be a part of a new Cunarder and her history from the very beginning was truly special.

Although I had previously worked on the new-build team of four ships, when I first saw *Queen Victoria* in the Fincantieri shipyard in July 2007, I knew this was going to be not only very different to what I had experienced before, but also momentous.

Looking around the ship that day, you could feel a whole new level of pride, even from the shipyard workers who would have been involved with numerous ship deliveries beforehand. As you will see in Chris and Rachelle's stunning pictures, so much thought has gone in to the design of *Queen Victoria*, from the references to her namesake, to the Cunard signatures such as the Queens Room and the Golden Lion Pub. Cunard loves celebrating tradition and the company's history, but they also like to introduce new ideas like Cunardia, the first museum at sea. Then, of course, we were also the first ship to offer guests fencing classes in our sports programme. One of the greatest pleasures for me as an Entertainment Director was to hear the expressions of surprise and delight when guests saw our magnificent three-deck-high Royal Court Theatre, with the first theatre boxes at sea.

(Courtesy of Vicki Cross)

I was lucky to go back to the shipyard near Venice on a few occasions, for press visits and to meet the Royal Household when we discussed the naming day's arrangements. On each visit the progress was remarkable, and to see the extraordinary talent and professionalism go in to every aspect of the ship's construction was a real treat. I have enjoyed so many memorable days on *Queen Victoria*, but when Captain Wright announced to the crew, on

20 November 2007, that she had been officially handed over to Cunard, we all knew it was another great day in Cunard's history. This was only topped nine days later by the magnificent naming ceremony in Southampton with her Royal Highness, The Duchess of Cornwall, in the presence of His Royal Highness, Prince Charles. Despite all the preparation and hard work that went in to preparing *Queen Victoria* for her career at sea, the real test is when our guests

come on board and how much they enjoy their voyage. We are fortunate that Cunard has such a loyal following, with so many guests travelling with us time and time again. We knew that guests may take a little time to adjust sailing on a new Cunarder, and that *Queen Victoria* would have to earn her stripes as it were. I think we were all overwhelmed by the enthusiastic response from all our guests on that very first voyage to Northern Europe, and then on her maiden world voyage. What a great reward for the crew on board who had worked incredibly hard to ease *Queen Victoria* in to service. The teamwork involved was a true testament to the Cunard spirit, and marked the beginning of the now familiar phrase 'We are Cunard', which is a way we express our pride in our ships and, of course, our legendary White Star Service.

I hope you feel the unique atmosphere of *Queen Victoria* through these wonderful images and understand why she has such a special place in our hearts. It is a wonderful tribute to a remarkable Cunarder, who, I am sure you will agree, is a just holder of the title 'Queen'.

Alastair Greener

2010

(Courtesy of Alastair Greener)

ACKNOWLEDGEMENTS

We would like to thank everyone who helped us create our photographic journey aboard *Queen Victoria*.

Special thanks goes to:

Captain Christopher Rynd, for his hospitality aboard and for writing his perspective of being master of *Queen Victoria*; Entertainment Director, **Alastair Greener**, for writing the foreword and for his company aboard the ship; Social Hostess, **Jennifer Schaper**, for organising the tours of the ship's various behind the scenes areas; **Chief Engineer Martyn Elliot**, for the comprehensive tour of the Engine Room; **Executive Chef Nicholas Oldroyd**, for the fantastic tour of the ship's galley, store rooms and crew mess; **Third Officer Jack Martin**, for his insight into the ship's Bridge, **Youth Director Paul Trotter**, for organising access to The Zone; and **Valentin Angelov**, for organising access to the Mauretania Suite.

We are extremely grateful to **Amy Rigg**, **Emily Locke**, **Glad Stockdale**, **Ben Parker**, **Jennie Younger** and everyone at The History Press for their ongoing support for this, the third book in our 'A Photographic Journey' series.

Our thanks to **Caroline Matheson**, **Tim Wilkin** and **Julia Young** of the Cunard Insights team, for their ongoing support of our maritime lecturing endeavours; **Michael Gallagher** from the Cunard Line for his support in accessing Cunard images of *Queen Victoria*'s life so far; and to **Bill Miller**, for is continued support and photographic assistance.

Our thanks also to **Andrew Sassoli-Walker**, **Andy Fitzsimmons**, **Alan Gould**, **Alex Lucas**, **Debora Silva**, **Henryk Matysiak**, **James Griffiths**, **Rob O'Brien**, **Rosie Claxton**, **Russ Willoughby** and **Vicki Cross** for photographic or material assistance, and Casino Bartender **Martin G. Altares Jr**, for posing in our Casino photographs, and to our families for supporting us.

All photographs, unless otherwise credited, were taken by Chris Frame or Rachelle Cross.

INTRODUCTION

The *Queen Victoria* offers a timeless cruising experience, blending the classic vintage of a Cunard voyage with the modern amenities of a state-of-the-art cruise ship. Beautifully decorated throughout in a Victorian theme, you could be forgiven for imagining yourself travelling on an historic liner of the 1900s. In addition to the traditional luxury, you will also find on-board all of the amenities and conveniences that make twenty-first-century ocean travel so rewarding.

The ship's state-of-the-art podded propulsion system offers great manoeuvrability, allowing *Queen Victoria* to visit ports that her forebears were unable to access. This, together with her strengthened hull and stability system, provides guests with a thoroughly comfortable cruise to varied and interesting destinations worldwide.

Matching her gracious interiors, the time-honoured White Star Service on board ensures a level of attention that mirrors that found aboard White Star Liners such as *Olympic* and *Majestic*. Indeed, as Cunard today is associated with the finest luxury voyage, so too was White Star Line in the golden age of the ocean liner.

With itineraries that span the globe, *Queen Victoria* offers passengers not just a chance to relax and recuperate from their often hectic everyday lives, but also the opportunity to explore the world.

The Cunard livery, an international travel icon since the 1840s, has made *Queen Victoria* an instant celebrity. She is welcomed warmly in the ports that she visits. Though a young ship, the pleasing blend of awe-inspiring grandeur and intimate hideaway bars and lounges means that *Queen Victoria* has already won the hearts and minds of many repeat guests.

LOOKING BACK

For over 170 years, the name Cunard has been synonymous with transatlantic ocean liners. The company, founded by Sir Samuel Cunard in 1839, gave birth to the first regular transatlantic passenger crossings off the back of a British Government-funded mail service. Operated by a fleet of paddle steamers, the first Cunarders begun a tradition of British-built ocean liners that were known worldwide for their speed, grace and safety.

Names such as *Lusitania*, *Mauretania*, *Aquitania*, and of course the Queens – *Elizabeth* and *Mary*, were recognised across the globe as being the finest examples of ocean liner architecture.

However, despite their dominance on the North Atlantic, Cunard were no strangers to sending their ships cruising. In fact, during the Great Depression, Cunard were forced to paint their once mighty *Mauretania* in a white livery and send her cruising to the tropics in order to make ends meet. Despite cruising duty during the Depression, it was not until the 1940s that Cunard opted to venture into the cruise market full time. Built at the famed John Brown shipyard in Scotland, *Caronia* was Cunard's first purpose-built cruise ship, and she was quite different from any Cunarder before her.

Caronia sported a single mast and a huge single stack (the largest on any ship at the time), giving her a distinctive silhouette. Furthermore, she was painted in four shades of green, which earned her the nickname of the Green Goddess and made her instantly recognisable worldwide. She had an outdoor pool, sun-bathing decks and a more casual décor than her fleet mates, making her suitable for long holiday voyages. Despite her role as a cruise ship, *Caronia* was also able to travel the North Atlantic, and undertook several crossings for her owners during her life. This design – part cruise ship, part Atlantic liner – became known as dual purpose.

The most famous dual-purpose liner in history is the legendary *Queen Elizabeth 2*. More recently, the *Queen Mary 2* has risen to fame as the largest dual-purpose liner ever built, and her design follows the same basic principles laid down by *Caronia*. It is a delicate balance between the strength, power and stability required to plough the North Atlantic, while still offering the amenities that are expected when the ship goes cruising, such as pools, spas, nightclubs and balcony accommodation.

In the 1980s and 1990s, the tourism industry experienced a cruising revolution. New companies started to appear in the Caribbean, and with them a new breed of ship was coming of age. The modern cruise ship is a far cry from the traditional Atlantic liner. In fact, it is difficult to draw many comparisons between it and a classic cruise ship such as *Caronia*.

Above right: Cunard Princess. **(Courtesy of Russ Willoughby)**

Today's cruise ship designers focus on maximising passenger areas. Knowing their ships will be restricted to the relatively calm waters of the Caribbean or Mediterranean, they have sacrificed the long bows, deep draughts and swept superstructures of the traditional liner to make way for row after row of balcony accommodation.

By the late 1990s, Cunard was feeling the pressure from increasing competition. Reliant on the famed *QE2* and a fleet of small cruise ships such as *Cunard Countess* and *Cunard Princess*, their aging fleet was unable to match the economies of scale or creature comforts of the latest new-builds. Fortunately for the historic company, the Carnival Corporation purchased it in 1998 and set about repositioning Cunard as a luxury brand. *QE2* was complemented with a big sister, *QM2*, and plans were announced for the construction of a third Cunard Queen.

Queen Victoria was originally to be delivered in 2005, however, after the success of *QM2*, Cunard's design team went back to the drawing board. The ship originally intended as *Queen*

Victoria was christened *Arcadia* for P&O Cruises (also owned by Carnival), while the yard number 6127 was allocated to the new Cunarder at the Fincantieri shipyard in Italy.

Although one could be forgiven for not noticing the differences between *Queen Victoria* and *Arcadia*, given that they are both based on Carnival's Vista class design, behind the scenes number 6127 was altered, making her by far the strongest ship in her class. Her bow was significantly strengthened while her decks were rearranged, allowing for an interior layout reminiscent of the great Cunarders of days gone by.

What resulted was a cruise ship like no other. The first Cunarder built in Italy would be able to transit the North Atlantic Ocean in seven days, making her the first cruise ship designed with the direct Atlantic passage in mind. In fact, during her maiden Atlantic crossing in January 2008, when she sailed in tandem with *QE2*, she was met by severe weather yet escaped unharmed.

Queen Victoria is a proud custodian of the Cunard name, and ensures that the future of Sir Samuel's legacy will survive for decades to come.

WELCOME ABOARD

From floor to ceiling, *Queen Victoria*'s Grand Lobby is designed to impress. Situated over three levels, the room commands attention with its sheer size as well as its elegant decor. A grand staircase, reminiscent of those aboard the White Star Line's historic Olympic class liners, connects Deck 1 and Deck 2. On the other side of the lobby, dual curved staircases link Deck 2 and Deck 3.

Above the grand staircase is an art deco-style arch with a bas-relief of the *Queen Victoria* at sea, created by artist John McKenna, who also created the bas-relief in the *QM2*'s Grand Lobby.

A dome-shaped three-tier chandelier is centred over the lobby and provides that final touch of glamour. There are many rooms located off the Grand Lobby, ensuring that passengers have plenty of opportunities to enjoy this space.

DID YOU KNOW?

Queen Victoria is the fifth Cunard Queen.

ACCOMMODATION ABOARD

A comfortable cabin is an expectation of today's cruise ships, and *Queen Victoria* does not disappoint. Her accommodation is graded by luxury and ranges from relatively spacious inside rooms to the very deluxe Mauretania, Laconia, Aquitania and Berengaria suites occupying the aft space on Deck 6 and Deck 7. Each cabin is kept spotless by a team of cabin stewards and butlers who clean all rooms twice daily and are only a phone call away should you require anything. Standard across all cabin grades is a nightly turn down service, 24-hour room service, regularly replenished toiletries and a daily shipboard newspaper to keep you apprised of the news at home.

QUEENS GRILL ACCOMMODATION

For those travelling Queens Grill grade, the ultimate in luxury awaits. With the Queens Grill grade reserved for the biggest and best of *Queen Victoria*'s suites, it is little wonder that these staterooms carry the highest price tag. Those who can afford it will enjoy the decadence of a personal butler, fresh flowers, a well-stocked mini bar and pre-dinner canapés.

For those who really want to spoil themselves, the Mauretania, Laconia, Aquitania and Berengaria suites, on the aft of Decks 6 and 7, are the epitome of Cunard luxury.

PRINCESS GRILL ACCOMMODATION

The *Queen Victoria*'s Princess Grill accommodation provides its guests with a concierge and priority embarkation service, in addition to the services provided to Britannia passengers. All of the Princess Grill suites have a balcony and a sitting area in their cabin. Passengers travelling in Princess Grill accommodation grade are given access to the Grills Lounge and the Terrace, which are exclusive Grills only areas.

BRITANNIA STATEROOMS

Britannia accommodation encompasses the majority of cabins aboard *Queen Victoria*, and as such provides a wide variety of options. Britannia cabins include inside cabins with no natural light source, ocean view cabins with windows, and balcony cabins. Prices range over the categories, and the cabins are distributed over all passenger accommodation decks.

Passengers staying in the Britannia level of accommodation dine in the Britannia Restaurant and have access to most passenger areas.

On the 24-hour room service:

'Is it available after hours?'

RESTAURANTS

Despite the many activities offered aboard *Queen Victoria*, somehow the food always manages to feature as a highlight of the day. With a wide range of dining options for every meal, and snacks and treats available throughout the day, passengers will never go to bed hungry.

In addition to providing round-the-clock dining options, the restaurants aboard *Queen Victoria* are designed to impress, each restaurant having a distinct look and feel. With a large collection of paintings, models and sculptures displayed in and around the restaurants, they have an appeal that is not limited to the fantastic meals they serve.

RESTAURANTS PROFILE

RESTAURANT	LOCATION	SEATINGS	CAPACITY
Queens Grill	Deck 11	Single Seating	130
Princess Grill	Deck 11	Single Seating	120
Britannia Restaurant	Decks 2 & 3	Double Seating	900
The Lido	Deck 9	Open Buffet	450
The Lido Pool Grill	Deck 9	Open Grill	Takeaway
Todd English	Deck 2	Single Seating	100

QUEENS GRILL

The Queens Grill restaurant, on the port side on Deck 11, is designated for the use of passengers travelling in the Queens Grill class accommodation. The Queens Grill has windows along the outer wall which curve out over the sides of the ship, providing a truly spectacular view.

The Queens Grill opens out onto the Courtyard, where Grill-class passengers can dine outdoors under the shade of umbrellas.

Featuring stained-glass windows and a number of large murals, this restaurant is lit by lamps and large chandeliers as well as subtle back lighting, giving the elegant atmosphere one would expect from a Cunard Grill.

DID YOU KNOW?

The first Queens Grill restaurant aboard a Cunard ship was the one added to *QE2* in 1972.

PRINCESS GRILL

Mirroring the Queens Grill on Deck 11 is the Princess Grill restaurant. Like the Queens Grill, the Princess Grill has curved windows along the outer edge of the ship, offering an endless view to the horizon. The restaurant's décor reflects that of the Queens Grill, with subtle lighting and neutral furniture, offering classic British understated luxury.

Princess Grill guests can also access the Courtyard and dine under the stars, a popular option on warm summer cruises.

BRITANNIA RESTAURANT

The largest of the formal dining rooms, the Britannia Restaurant occupies the space on the aft of Decks 2 and 3. The Britannia Restaurant caters to 70 per cent of the passengers on board, and although it is a double-height room, two seatings are necessary to accommodate the Britannia passengers at dinner time. Seating is allocated in the evenings, but there are no such restrictions at breakfast and lunch times.

The Britannia Restaurant is a visually stunning room decorated in autumnal tones. Its focal point is a large bronze globe, which is displayed in front of a freestanding double-height

mural. The upper storey of the restaurant is open in the middle, with a series of gracefully curving balconies overlooking the lower floor.

Decorated in an art-deco theme, the Britannia Restaurant was inspired by the dining carriages of the *Orient Express*. With large windows flanking the port and starboard sides, this restaurant has the impressive feature of windows overlooking the stern, offering superb views of her wake as *Queen Victoria* makes her way from port to port.

DID YOU KNOW?

The Britannia Restaurant takes its name from Cunard's first ship.

LIDO

For those who enjoy the choice of a buffet and a casual dining atmosphere, the Lido restaurant on Deck 9 is the place to go. The buffet is arranged so that passengers have the chance to peruse the entire selection of food on offer, with two separate buffet lines to reduce waiting times.

The Lido has a very relaxed feel, which makes it a popular alternative for many diners. Windows run the entire length of the room, providing a fantastic amount of light and allowing

FANCY A PIZZA?

One of the most popular items on the menu in the Lido is the homemade pizzas. Diners can order the pizza of the day, or make their own combination which is cooked fresh to order. Chefs will dazzle guests with their professional pizza-making abilities – including the seemingly risky throwing of the dough.

views of the endless blue of the ocean. The ceilings sport tinted mirrored surfaces, offering the illusion of even more space and height, further emphasising the spaciousness of this restaurant.

The seating is arranged mostly in tables for four following the edges of the room, but there are also tables for two and several tables with curved padded bench seating scattered about.

LIDO POOL GRILL

Located on Deck 10 between the Lido restaurant and the Lido Pool is the Lido Pool Grill. The Lido Pool Grill serves light grill dishes, including burgers, hot dogs and sandwiches from 11a.m. to 5p.m. The Lido Pool Grill allows those who are enjoying their time by the pool to eat without abandoning their deck chair.

TODD ENGLISH

On the port side on Deck 2 you will find the alternative dining venue, the Todd English Restaurant. Open for lunch and dinner, Todd English serves a Mediterranean menu that was designed by its celebrity chef namesake.

Some accommodation packages include a meal at this restaurant, but for most there is a tariff charged for this experience. With its excellent reputation and unique menu, the restaurant is often full, and a table must be booked in advance.

Todd English is decorated in a bright colour palate, with gold and burgundy carpeting complemented by light drapes that are illuminated from the top to give an elegant evening atmosphere. Red vases further enhance the room, offering additional light by way of a globe inside each, and are placed about the room on dark wooden side tables.

BARS AND LOUNGES

I f you fancy a drink, or would simply like a comfortable place to sit and read, you will find a wide range of bars and lounges aboard to suit your needs. In the mood for a summery strawberry daiquiri? Then head on deck to the Lido Bar. If you would prefer a pint of Guinness to go with your traditional fish and chips lunch, make your way to the Golden Lion. Or, if it is a bit early in the day, you may prefer to simply sit by a window in Café Carinthia and enjoy a cafe latte as you watch the world go by.

FANCY A DRINK?

Queen Victoria's passengers will consume annually:

Champagne/Sparkling:	119,400 bottles
Red Wine:	109,000 bottles
White Wine:	119,600 bottles
Dessert Wine:	3,900 bottles

BARS AND LOUNGES PROFILE

BARS AND LOUNGES	LOCATION
The Grills Lounge	Deck 11
Commodore Club	Deck 10
The Admirals Lounge	Deck 10
Churchill's Cigar Lounge	Deck 10
Hemispheres	Deck 10
Pavilion Bar	Deck 9
Winter Garden	Deck 9
Lido Bar	Deck 9
The Alcove	Deck 3
Midships Lounge	Deck 3
Card Room	Deck 3
Casino	Deck 2
Golden Lion	Deck 2
Queens Room	Deck 2
Champagne Bar	Deck 2
Café Carinthia	Deck 2
Chart Room	Deck 2
Royal Court Theatre	Decks 1, 2 & 3

Joke overheard in the Chart Room bar:

'How can you tell that a passenger has had too much to drink?'

'They are the only one able to walk in a straight line down the corridor!'

THE GRILLS LOUNGE

Those passengers who travel in the Grills class will have exclusive access to The Grills Lounge. The ambiance here is formal, with a décor that hints of upper-class elegance. The highest lounge aboard the ship, and situated between the Queens and Princess Grill restaurants, the lounge is dominated by a domed stained-glass ceiling that bathes occupants in soft light.

Grill-class passengers will find their personal concierge here, who is on hand to provide the attentive service one would expect from the Grills experience. Its location, close to the restaurants, makes this room popular for pre-dinner drinks, while others prefer to adjourn here for a nightcap.

COMMODORE CLUB

Continuing the Cunard tradition of a forward-facing observation lounge, the Commodore Club aboard *Queen Victoria* offers sweeping forward-facing views that are second to none. With a name that pays tribute to the same room aboard *Queen Mary 2*, the Commodore Club aboard *Queen Victoria* is located on Deck 9 and is accessible from the A Stairway.

By day a quiet lounge, the bar is a popular place to sit and watch the world go by; by night the room transforms into a cocktail bar with a sophisticated atmosphere. Situated above the Bridge, it is the highest indoor vantage point aboard *Queen Victoria* that offers a forward view and is a favourite amongst passengers.

THE ADMIRAL'S LOUNGE

A hallmark of all great Cunarders is the inclusion of small hideaway bars and lounges that surprise and delight those inquisitive enough to find them. The Admiral's Lounge is one such place aboard *Queen Victoria*.

Nestled between the Commodore Club and Churchill's Cigar Lounge, this quiet space is seldom used for scheduled functions and offers guests a private place to meet with friends. Book lovers will enjoy the area for its tranquillity, and, as such, readers are often found in this comfortable retreat.

CHURCHILL'S CIGAR LOUNGE

Refined and elegant are two watchwords of the Churchill's Cigar Lounge. Located on Deck 9 near the Commodore Club, the lounge offers pleasing ocean views for those passengers who fancy a cigar.

The room is decorated in green hues with warm wood panelling and leather chairs, giving it an authentic appearance.

DID YOU KNOW?

During the Second World War, Sir Winston Churchill sailed a number of times between Britain and America aboard Cunard's *Queen Mary*. After the war, Churchill credited the Cunard Queens with assisting in shortening the war.

HEMISPHERES

Hemispheres is *Queen Victoria*'s nightclub. Located on Deck 10, the room is curved with a circular dance floor located in the centre under a raised ceiling. The bar itself is also curved, as is much of the artwork in the room. With neon blue lighting over the dance floor the room is well equipped for those who wish to dance the night away. By day, this bar is a comfortable, quiet place to enjoy a drink or a chat.

PAVILION BAR

Thirsty passengers lounging by the Pavilion Pool will appreciate the Pavilion Bar. Located at the forward end of the pool area, this bar is kept busy throughout the day by eager swimmers and sunbathers alike. Accessed from either the A Stairway (forward) or the Winter Garden (aft), this bar is close to magnificent ocean views thanks to large glass screens that run lengthwise on both the port and starboard side of the nearby Pavilion Pool. Expert bartenders will tend to your every whim here, with cocktails being a favourite during summer cruises.

THE VERY POPULAR STRAWBERRY DAIQUIRI

50ml of white rum
10ml freshly squeezed lime juice
5 ripe strawberries
Sugar syrup

Blend (or shake) ingredients along with 4 ice cubes, pour into a chilled margarita glass and garnish with a strawberry.

WINTER GARDEN

The Winter Garden offers a truly unique on-board ambiance. Wood panelling, wrought iron balustrades and arched windows suggest old world charm, however visitors to this room need only look up to realise they are aboard a twenty-first-century Cunarder.

A retractable glass roof, known as a Magrodome, covers the Winter Garden, allowing passengers to be draped in the natural warmth of the sun while enjoying the garden-like surroundings. Natural greenery is a prominent feature here, and, combined with the comfortable atmosphere and wicker furniture, results in this being a popular space aboard *Queen Victoria*.

LIDO BAR

Those with a passion for fresh air, swimming and a tipple of their favourite drink will love the Lido Bar. Set just forward of the Lido Pool, the Lido Bar is popular during the day due to its outdoor location. With an atmosphere of casual relaxation, the bar spills out onto the open deck that surrounds the Lido Pool. Here, passengers will enjoy expansive views of the ocean as well as finding *Queen Victoria*'s stern staff which proudly flies the Red Ensign.

DID YOU KNOW?

Queen Victoria is registered in Southampton and, as such, flies the Red Ensign.

THE ALCOVE

S andwiched between the upper floor of The Library and the Card Room, The Alcove is exactly what its name implies. The Alcove is the location of a model of the ship and two tables where passengers can contribute their skills to putting together the jigsaw puzzles that are placed here for that purpose.

Two large windows look out from The Alcove over the port side of the Promenade Deck and the room is open to the Grand Lobby, providing spectacular views on both sides.

MIDSHIPS BAR & LOUNGE

With an interior view overlooking the Grand Lobby, the Midships Bar & Lounge aboard *Queen Victoria* is the perfect place for people watching. Located amidships on the starboard side of Deck 3, the bar also boasts large windows providing views onto the Promenade Deck.

DID YOU KNOW?

Queen Mary of 1936 was originally intended to be called *Queen Victoria*.

CARD ROOM

You could be forgiven for thinking you have stepped back in time when entering the *Queen Victoria*'s Card Room. Decorated in dark wood panelling and featuring a period ceiling, the room is reminiscent of similar spaces found aboard Atlantic icons such as the likes of *Mauretania* and *Olympic*. The Card Room is the perfect place to brush up on your gaming skills, with bridge lessons being a popular pastime in this room. Located on the port side of Deck 3, the Card Room is centrally positioned, just a short walk from the upper level of the Library.

EMPIRE CASINO

Whether you are feeling lucky or are an expert at the game, you will enjoy a visit to the Empire Casino. Located on Deck 2, the Empire Casino is decorated in a style reminiscent of Cunarders from a bygone era. Backlit stained-glass ceilings are set atop stylised pillars, creating the perfect setting for an evening of refined entertainment.

While some guests like to try their luck on the poker machines, others prefer to get together as a group and play one of the many table games, including roulette, craps, blackjack and poker.

If you are a novice, or simply wish to brush up on your skills, expert croupiers are on hand to teach passengers how to play the various games – but do not get too comfortable, as the house often wins!

Queen Victoria Empire Casino
bartender Martin G. Altares Jr.

GOLDEN LION

Though a later addition to the *QE2*, the Golden Lion Pub proved so popular that this room has been featured on each of Cunard's modern Queens. *Queen Victoria*'s Golden Lion takes the British-pub-at-sea experience to a new level, with a Victorian-style glass frontage featuring period lamps. Bay windows set within the hull look out over the ocean on the starboard side. The Golden Lion is fitted out

with dark polished wood furniture, green and dark-red leather. Lighting in the Golden Lion is dim, enhancing the atmosphere of the room.

The Golden Lion is a popular location for lunches on sea days. The pub serves traditional British favourites including steak and mushroom pie, fish and chips and a ploughman's lunch. This bar serves beer and ale on tap, but other drinks are also available.

THE QUEENS ROOM

DID YOU KNOW?

The Queens Room is based on Queen Victoria's favourite residence, Osborne House.

Grand style and elegance reminiscent of a bygone era can be found in *Queen Victoria*'s Queens Room. Situated on Deck 2, the room boasts an impressive double height with balconies, accessible from Deck 3, offering passengers a unique perspective of this popular room.

While many passengers will remember this room for the ballroom dancing, cocktail parties or classical concerts, most will recall visiting for the Cunard tradition of afternoon tea, served daily at 3.30p.m.

QUEENS ARCADE

Running along the starboard side of the Queens Room is the Queens Arcade. Separated from the corridor by a low balustrade, the Queens Arcade offers tables with a view of the ocean and the Queens Room.

Comfortable seating is available here, and it is a fantastic location to take advantage of the daily high tea ritual.

THE CUNARD WORLD CLUB PARTY

If you are a past passenger of the Cunard Line, you will be part of the Cunard World Club. The loyalty programme offers guests a number of benefits, none of which are better received than the Cunard World Club Party which is held aboard during each cruise.

The party is typically hosted in the Queens Room and guests are treated to complimentary cocktails, champagne and beer (as well as soft drinks). Canapés are served while passengers mingle. The highlight of the event is a speech by the Captain, when the senior staff are introduced. Guests will also learn just how many voyages their fellow passengers have taken, with the top sailing passengers for that voyage being congratulated.

DID YOU KNOW?

True to her Victorian heritage, *Queen Victoria* offers fencing lessons aboard (the first Cunarder to do so), which are held in the Queens Room.

CHAMPAGNE BAR

The champagne bar offers guests a treat with an impressive selection from the Veuve Cliquot cellar. Located off the Grand Lobby on Deck 3, the bar is decorated in an art-deco theme, including murals that pay homage to one of the greatest Cunarders of all time, the *Queen Mary*. The central circular bar is surrounded by

comfortable lounge seating. While a popular location for catching ocean glimpses during the day, at night the bar comes alive as guests enjoy a glass of the finest champagne in period surroundings.

CAFÉ CARINTHIA

Café Carinthia is a popular lounge and bar and is in use from early morning until late at night. Serving specialty teas, coffee and pastries from morning to afternoon, Café Carinthia is an enjoyable place to spend a few hours with a good book or new friends.

At night it is a popular place for a pre-dinner drink, and the lights are slowly dimmed to provide an intimate bar experience. The room has dark wood finishes and heavy drapes on the windows, making it feel luxurious and elegant.

CHART ROOM

A s one might expect, given its name, the Chart Room aboard *Queen Victoria* sports a nautical theme. Traditional maritime artwork is juxtaposed with a more modern take on 'the chart', with maps superimposed on mirrored backdrops, creating an impressive visual effect of modern elegance.

Situated close to the Britannia Restaurant and offering an atmosphere reminiscent to that of a private club, this attractive space is quick to fill in the evenings as passengers relax with a pre-dinner drink and listen to the live musical performances.

ROYAL COURT THEATRE

Named for the room found aboard the Cunard Flagship, *QM2*, the Royal Court Theatre aboard *Queen Victoria* serves as the ship's primary show lounge. The room, located over three decks (Decks 1, 2 and 3), offers two levels of box seating, the first aboard a cruise ship.

A large mechanised stage is the centrepiece of this room, allowing the Royal Cunard

Singers & Dancers to perform a number of elaborate shows during the cruise. Backed by an impressive state-of-the-art sound and lights system, passengers will more often than not recall the shows as one of their favourite pastimes aboard *Queen Victoria*.

In addition to being a show lounge, the Royal Court Theatre is also the location of the enrichment lectures given by the guest speakers who travel on board for this purpose.

PUBLIC ROOMS

On occasions when you are not eating, drinking or sleeping, the public rooms on *Queen Victoria* offer exciting spaces to explore. With the opportunity to work out your body, your mind and your credit card, guests can have a day as relaxing or as busy as they please.

Queen Victoria's public rooms are primarily based on Decks 1, 2 and 3, with the Grand Lobby being the central focal point. From here, passengers can visit a number of areas that allow for a variety of activities. From trying your luck at the Casino to working out at the Royal Cunard Gym, there are pastimes to suit every personality.

PROFILE OF PUBLIC ROOMS

ROOM NAME	LOCATION	POPULAR FOR
The Zone	Deck 10	Video games (Teens Only)
The Play Zone	Deck 10	Entertaining the kids
Cunard Health Club & Spa	Deck 9	Keeping healthy during your voyage
Royal Arcade	Deck 3	Shopping!
Art Gallery	Deck 3	Art connoisseurs to browse or buy
ConneXions	Decks 3 & 1	Enrichment and Internet Access
Cunardia	Deck 3	Historical Cunard information
The Book Shop	Deck 3	Buying maritime books and souvenirs
Images	Deck 3	Checking out last night's photos!
Library	Decks 3 & 2	Finding something to read
Internet Centre	Deck 1	Browsing the web, sending emails
Tour Office	Deck 1	Organising shore tours
Pursers Office	Deck 1	General reception and concierge services

THE PLAY ZONE

Midships on Deck 10 are the dedicated under 18s' areas. The Play Zone caters to children and offers age-specific entertainment and a way to meet new friends. The staff of The Play Zone organise activities and games for younger passengers, as well as providing supervision for children to allow parents to have some time to themselves.

The Play Zone opens out on to a private outdoor area, not accessible to other passengers.

THE ZONE

Teenagers (under 18) will love The Zone. A hive of modern activities, The Zone has everything from Air Hockey to PS3s!

Overseen by youthful cruise staff members, The Zone has a fun, relaxed atmosphere and is the perfect escape for younger cruisers wanting to take a break from the formality of shipboard life.

CUNARD ROYAL SPA AND FITNESS CENTRE

The Cunard Royal Spa and Fitness Centre, on the forward end of Deck 9, offer the opportunity for passengers to indulge their senses and increase their wellbeing. The fitness area features cardio equipment, weights and an aerobic area, with classes offered by the dedicated gym staff.

The Spa offers a thalassotherapy pool, thermal suites, a hairdresser and a beautician. Passengers can partake of a massage and alternative therapies for the ultimate in relaxation.

ROYAL ARCADE

A spectacular shopping experience can be found on Deck 3 at the Royal Arcade. Located just aft of the Royal Court Theatre, passengers will enjoy an engaging shopping experience here, whether they are serious buyers or simply window shoppers.

The Arcade, inspired by both the Royal and Burlington Arcades of London, has a double curved staircase, which surrounds an ornate clock designed to resemble the Queen's Arcade clock (also from London).

Heard at the Royal Arcade:

'What time is the Captain's noon report?'

With wrought iron balustrades and nineteenth-century-style lighting, this two-level arcade houses a variety of merchandise including jewellery, perfume, formal-wear and Cunard-branded apparel.

For those who enjoy bargain hunting there are regular sales held in this area, with sale items laid out on trestle tables, creating a market atmosphere.

ART GALLERY

Those interested in art and design will enjoy wandering the corridors of *Queen Victoria*, but if you are interested in a work of art to take home with you then the Art Gallery is the place to go.

The Art Gallery, located starboard on Deck 3, is where you can go to view paintings, sculptures and other art pieces that are carried aboard *Queen Victoria* for sale. These items are available for viewing throughout the trip, and art auctions are held every few days to enable passengers to bid for their favourite works.

CONNE X IONS

The ConneXions rooms are located on Decks 1 and 3. The Deck 1 ConneXions room is fitted out with computers and serves as a secondary Internet access point. The Deck 3 Connexions facilities are the location for meetings, small lectures and some administrative operations. For most of the time they are multi-purpose rooms, used as and when required.

CUNARDIA GALLERY

The Cunard Line's history dates back to 1839, when Sir Samuel Cunard travelled to England to secure the British Royal Mail contract to provide a regular scheduled mail service across the North Atlantic. This event, and many more, are depicted aboard *Queen Victoria* in the Cunardia Gallery.

There are several areas aboard *Queen Victoria* with Cunardia information, but the main gallery is located on Deck 3, just aft of the Royal Arcade.

The Cunardia Gallery features photos, prints and paintings which cover some of the more interesting aspects of Cunard Line's long history.

THE BOOK SHOP

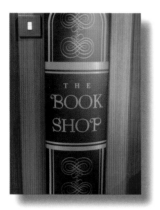

The Book Shop on Deck 3 stocks a wide range of maritime souvenirs with a strong focus on Cunard ships, past and present. In addition to books passengers can browse the selection of playing cards, bookmarks, posters, pens and postcards.

The Book Shop also caters to passengers who are not so nautically inclined, with a selection of fiction and non-fiction books for sale.

IMAGES PHOTO GALLERY

Y ou will feel like a celebrity on a formal night aboard *Queen Victoria*. The ship has a team of photographers who capture various moments during your trip, with evening cocktail parties, formal dinner and elegance shots (taken near the Grand Lobby) being particularly popular.

The ship's photographers display their work in Images Photo Gallery, found starboard aft on Deck 3, near the upper level of the Britannia Restaurant. Photos purchased are presented in a Cunard-branded folder, though you can opt to purchase specialised leather folders or photo frames if desired.

LIBRARY

The *Queen Victoria*'s library is the kind of library that booklovers dream of. Dark wood, double height, with a spiral staircase providing access to both levels, this room is dramatic and sure to strike a chord in those with a taste for the written word.

The library has a selection of more than 6,000 books spanning a wide variety of genres, and guests can pass many happy hours browsing the shelves or relaxing in the comfortable chairs.

INTERNET CENTRE

No man is an island, not even a man at sea, and for passengers on the *Queen Victoria* there is no need to feel cut off from the rest of the world. The Internet Centre on Deck 1, just forward of the grand staircase on the port side, offers access to the Internet for a fee. Guests can use the computer terminals in this room, or can arrange access to the WiFi connection on the ship and use their own.

TOUR OFFICE

Whilst you may have come on board to holiday on the *Queen Victoria*, that does not mean that the ship is all you want to see. *Queen Victoria*'s cruise itineraries offer a chance for passengers to experience foreign shores, and the Tour Office staff are the people who help to arrange that. With numerous tours on offer in most ports, passengers can visit the Tour Office on Deck 1 to find out about, and book, their onshore adventures.

PURSER'S OFFICE

The administrative hub of the ship, the Purser's Office, is kept busy around the clock. Located on the starboard side of Deck 1, the staff of the Purser's Office keep things running smoothly. With responsibilities ranging from connecting telephone calls to compiling passenger accounts, most guests will find themselves in contact with the Purser's Office at some point during their cruise.

The Purser's Office has a curved marble desk, dark wood pillars and a warmly coloured mural on the rear wall.

(Courtesy of Alastair Greener)

ON DECK

There is no more relaxing feeling than lying back on a deckchair and watching the horizon pass by as *Queen Victoria* makes her way across the world's oceans.

Like all ships in the Cunard fleet, *Queen Victoria* has classic wooden steamer deckchairs, giving one the feeling of being on deck aboard a classic ocean liner. *Queen Victoria*'s promenade deck, situated on Deck 3, runs along both the port and starboard side of the ship and offers shaded outdoor space courtesy of the ship's sixteen lifeboats.

High atop the ship on Decks 9, 10 and 11, passengers can take in the sun in the ample outdoor space available. Whether relaxing with a book or enjoying an afternoon cocktail, you are sure to be content out on deck.

POPULAR ON-DECK SPOTS

AREA NAME	LOCATION	HERE YOU WILL FIND
The Terrace & The Courtyard	Decks 11 & 12	Exclusive space for Grills guests
Sports Deck	Deck 11	Paddle tennis courts, deck games
Pavilion Pool	Deck 9	Swimming Pool, Whirlpools, nearby bar
Lido Pool	Deck 9	Swimming Pool, Whirlpool, nearby grill
Promenade Deck	Deck 3	Deck chairs, superb views

THE TERRACE & THE COURTYARD

Passengers travelling Queens Grill and Princess Grill class will have exclusive access to the Terrace and the Courtyard on Decks 11 and 12.

The Courtyard is an alfresco eating area, located between the Queens Grill and Princess Grill restaurants. Passengers can opt to eat breakfast, lunch, afternoon tea or dinner here.

Forward of the Grills Lounge is the Terrace, a great place for guests to enjoy the outdoors whilst also enjoying a drink from the bar.

Upstairs on Deck 12 is the Grills Upper Terrace, where deckchairs are arranged so that Grills passengers can enjoy the sun.

SPORTS DECKS

H igh atop the ship you will find the Sports Decks. Boasting an impressive array of sporting options, you will find a golf driving range, paddle tennis and basketball courts, as well as the more traditional shipboard games of shuffleboard and deck quoits.

With sweeping forward-facing views, the Deck 11 sports area is a popular location to exercise, offering guests the opportunity to indulge at dinner and not go home feeling guilty!

PAVILION POOL

The Pavilion Pool is located on Deck 9, aft of the Cunard Royal Spa and Fitness Centre and forward of the Winter Garden. Large enough to complete laps, this pool is often in use early in the morning for the more fitness-oriented passengers.

Sheltered from the wind (due to the overhanging deck that flanks each side of this space), the deckchairs here are always in use.

The Pavilion Pool is a popular location for night-time deck parties, with music provided courtesy of the ship's Caribbean Band.

LIDO POOL

Located at the aft of Deck 9 is the Lido Pool and Whirlpools. An open deck location, the Lido Pool is a great place to soak up some sun when the weather is fine. The Lido Pool area has fantastic views over the back of the ship, and is a popular location for passengers to wave goodbye when departing ports.

PROMENADE DECK

Located on Deck 3, the Promenade Deck offers passengers a space to relax and watch the world go by. With authentic wooden steamer deckchairs (complete with comfortable cushioning), the Promenade Deck is sheltered from the sun by the overhead lifeboats.

The Queen Victoria's Promenade Deck is unique in that it wraps around the stern of the ship, offering an unobstructed view of the wake.

Joggers will enjoy using the Promenade Deck to keep fit, while others prefer to simply take a leisurely stroll or find a vantage point to watch *Queen Victoria* enter or leave a port.

PROFILE

L ike most Cunarders before her, *Queen Victoria* sports the traditional ocean liner livery. A matte black hull is juxtaposed against her white superstructure.

Topped by a *QE2*-style funnel – painted in the historic Cunard colours (which date back to 1840) – the overall scheme gives *Queen Victoria* a proud appearance reminiscent of the golden age of travel.

THE VISTA SISTERS

The *Queen Victoria* is a member of the Vista Class of ships. Originally a design created for Holland America Line (HAL), the Vista stable has been expanded to include ships of the P&O, Costa Cruises and Cunard fleets. All of these companies form part of the Carnival Corporation.

Queen Victoria was originally expected to be delivered in 2005, which would have made her the first Vista ship delivered to a line other than HAL. However, after Cunard's success with *QM2*, the ship originally intended to become *Queen Victoria* was transferred to P&O,

(Courtesy of Cunard Line)

becoming *Arcadia*. A new Vista hull was allocated to Cunard and was scheduled to enter service in 2007.

While retaining the basic Vista design, alterations were made to the new ship, including a strengthened hull that was stretched to 964ft. The extra strength allows *Queen Victoria* to undertake transatlantic crossings, usually off limits to cruise ships. An additional deck was also included, along with the reorganisation of her interior rooms allowing *Queen Victoria* to offer a unique Cunard on-board experience that differs from the other Vista sisters.

DID YOU KNOW?

The first Vista Class ship was Holland America's *Zuiderdam*.

(Courtesy of Vicki Cross)

THE FUNNEL

Since Cunard was acquired by the Carnival Corporation in 1998, a number of steps have been taken to ensure that some similarity exists between the ships in the fleet. One such step was to give all new-builds a funnel reminiscent of the one aboard the most famous Cunarder of all, the *QE2*.

Although by no means a carbon copy, *Queen Victoria*'s funnel follows the same basic design. The stack (painted black) is wrapped within the funnel cowling (painted red), both of which are flanked by a 'scoop' that directs air up and over the funnel, assisting in dispersing the smoke and soot away from the aft decks (and thus keeping the area soot-free for sunbathers).

THE MAST

Another similarity aboard all modern Cunarders is the mast. Again inspired by the structure aboard *QE2*, the mast offers an essential contribution to the navigation of the ship. Radar and radio transmitters are positioned on it, while the navigational flags are flown from here, advising other vessels when *Queen Victoria* is under the command of a pilot, hove to or anchored.

DID YOU KNOW?

Queen Victoria's official letters are GBQV.

BEHIND THE SCENES

Asked of the Captain at his cocktail party:

'If you're here, who is driving?!'

Away from the glamour of the passenger areas, *Queen Victoria* is a hive of activity. Behind the scenes the crew work 24 hours a day to ensure everything from air conditioning to room service runs smoothly.

A team of 900 people operate the ship. An extensive range of careers are available aboard the ship: engineers, plumbers, cleaners, mechanics, chefs, waiters, photographers – you name it, *Queen Victoria* has it.

Every so often, guests see glimpses of the work that goes on to keep the ship operating at the expected Cunard standards. During hours in port, paint crews are found touching up the ship's exterior, while in the small hours you will spot cleaners vacuuming the stairs, corridors and lounges to ensure that they are spotless for the day ahead.

THE BRIDGE

S et at the forward end of Deck 8 is *Queen Victoria*'s Bridge. The nerve centre of the ship, the Bridge not only offers the Captain and his Officers supreme views forward (for safe navigation), but also houses sophisticated navigation equipment which allows *Queen Victoria* to sail virtually unaided to any port in the world. While the principal activity in the Bridge is the safe navigation of the ship, the area is also used as the primary response location for any on-board emergencies.

Adjacent to the Bridge is the Chartroom, where charts are stored covering all the major oceans of the world. This, coupled with the electronic charts, allows *Queen Victoria* to easily achieve her role as Cunard's global cruise ship.

DID YOU KNOW?

Queen Victoria is the first Cunard Queen to have passenger accommodation on a level higher than the Bridge.

POWER PLANT

A s passengers enjoy the day-to-day life aboard *Queen Victoria*, little thought is given to the power that allows the ship to effortlessly transport them from port to port. However, below the waterline there is a hive of activity that controls everything from fresh water to the temperature in your cabin.

Mechanically speaking, *Queen Victoria* has six Wartsila Sulzer ZA40 engines consisting of four 16-cylinder and two 12-cylinder diesels. Each cylinder has a bore of 400mm with a piston stroke of 560mm, offering a combined power of 63.4MW.

The engines provide power to all areas of the ship, with a vast quantity of it used to drive the ship's two podded propellers. Manufactured by ABB (Finland), the pods resemble large outboard motors which are suspended under the ship's hull. Replacing the traditional propeller, the pods benefit from 360-degree rotation, and thus perform the duty of the rudder.

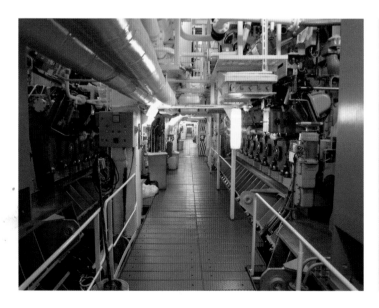

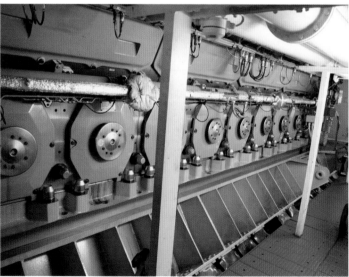

Despite the pods offering a superior level of manoeuvrability (compared to a traditionally powered ship), *Queen Victoria* further benefits from the inclusion of three bow thrusters. Located at the forward end of the hull, beneath the waterline, the bow thrusters have an output of 2.2MW each. This power, combined with the pods, allows *Queen Victoria* to dock without the aid of tugs.

THE STORES

Unlike many cruise ships operating today, nearly all food served on *Queen Victoria* is prepared fresh aboard. Everything, from the bread you eat at breakfast to the *petits fours* you have at the end of dinner, is created by the expert team in the ship's galleys.

As a result of this, the stores aboard *Queen Victoria* are immensely impressive. Fresh produce is ordered months in advance to ensure that passengers always get a varied

international menu. Main stores include red meats, fish, poultry, vegetables, fruits and dairy products (including the enviable chocolate larder).

One of the most valuable rooms per square metre is the wine and champagne store room. Here you will find one of the most varied cellars at sea, with an international range of red, white and rosé wine and, of course, champagne.

THE GALLEY

G uests dining in the restaurants aboard could seldom anticipate the size and scale of *Queen Victoria*'s galleys. In operation 24 hours a day, 7 days a week, there are various galleys aboard including the Britannia (lower and upper), Grills, Todd English and Lido.

Due to the size of the restaurant it feeds, the Britannia Galley is possibly the most impressive logistical success. The galley is organised into areas that work in unison to ensure that

waiters can collect orders in a straightforward manner, ensuring service is kept timely and food remains fresh and hot.

Over 100 Chefs work here under the direction of the Executive Chef. Once per voyage, passengers are invited to congratulate the team during the chef's parade, which takes place on the lower level of the Britannia Restaurant.

DID YOU KNOW?

The Britannia Galley can complete over 800 covers an hour!

CREW ONLY

The *Queen Victoria* carries with her a team of over 900 people, who work tirelessly at ensuring every passenger expectation is met. What may come as an unexpected surprise to passengers is the extent of the crew only areas aboard.

Crew areas have improved significantly in recent years, and *Queen Victoria* is no exception. There is a buffet-style eatery (called the Crew Mess) where the ship's crew are fed. The menu is varied, and includes both western and eastern dishes to suit the international team.

The crew bar, called The Pig & Whistle, is a Cunard institution and is a popular place to meet for a well-earned drink after work. Officers will enjoy the relaxed and refined atmosphere of the Wardroom, where historic shipping memorabilia forms a nautical theme.

There is a gym, a sunbathing deck and an outdoor jacuzzi for the crew to enjoy during their hard-earned down time, while others may prefer to use the crew-only WiFi Internet connection to stay in touch with loved ones at home.

DID YOU KNOW?

All of the Cunard Queens have a crew bar called The Pig & Whistle.

Although *Queen Victoria* was modified to allow her to undertake the occasional transatlantic crossing, she is first and foremost a cruise ship. She offers passengers a supreme level of creature comforts and a wide array of amenities suitable for cruising.

With a high number of outside cabins (most of which have a balcony), and a slower cruising speed than the previous Queens, the ship is perfectly suited for the slower pace of the pleasure voyage, and spends much of her time sailing the Mediterranean (both eastern and western).

Queen Victoria has also taken on the role of a global cruise ship, completing her maiden world cruise in 2008. During this voyage, she made several historic rendezvous with the outgoing *QE2* (which has since retired to Dubai), attracting large crowds eager to see the two Queens in such close quarters.

CAPTAIN'S PERSPECTIVE

One of the features that makes *Queen Victoria* distinctive is that she follows in the tradition of an ocean liner from the great age of ocean travel, and yet she a is a very modern vessel in terms of her external design, passenger facilities and technology. To sail on her is to experience the atmosphere, service and elegant lifestyle true to her heritage as a British ocean liner, yet within a contemporary hull.

Her interior layout and décor is shown well within the pages of this book: the Grand Lobby, the Queens Room, which is by the way one of the most popular places on board (serving as a ballroom with several themed balls every voyage), the traditional afternoon tea as well as Captain's cocktail parties.

There is the Library, a beautiful haven of peace and enrichment, the authentic Golden Lion pub, with all the atmosphere a good pub should offer (along with food and drink). The Royal Court Theatre hosts shows in the evening and lectures in the day from many notable experts in their fields of endeavor, who speak on a wide range of subjects to culturally enrich and inspire.

In terms of navigation and propulsion, she is a pleasure to handle. Her Azipod propulsion improves manoeuvrability, which enables visiting some smaller ports quite restricted in room and thus enables more options.

Technologically she is very modern and efficient in areas as diverse as storing provisions, fuel efficiency and in recycling and processing the various waste streams – all important considerations.

Spaciousness deserves a place and the choices it gives. Due to *Queen Victoria*'s relatively low number of passengers for her size, there is more room per person. This can be experienced on a day at sea in warm weather – *Queen Victoria* has so much deck space that

Above left: (Courtesy of Captain Rynd)

crowding is never an issue. Open decks can be enjoyed from a wooden steamer deckchair in the shade on the Promenade Deck, on one of the upper deck loungers, or, for the active, on the sports deck right forward on Deck 11.

The Grills restaurants are just superb, located high but centrally in the ship, offering breathtaking views and wonderful food and service. Our Grills passengers enjoy a higher level of luxury and service, an experience to aspire to.

As a Cunard liner we have signatures, being the latest embodiment of the legend, carrying on the traditions and legacy of 170 years of ocean liner travel, and elegant White Star Service from her loyal ship's company, whilst engaged in voyages worldwide. All of this makes for a memorable experience.

To have such a command is a privilege and a pleasure, but one more thing makes it unique for me and all of us on board – the sense of community. Being part of a small fleet, our ship's company, that is to say the officers, crew and our guests, develop a sense of community, of being part of the same enterprise. That is the unique difference.

Christopher Rynd

Captain, *Queen Victoria*

March, 2010

QUEEN VICTORIA FACTS

Gross Registered Tonnage:	90,000 GRT
Length:	964.5ft/294m
Beam:	106ft/32.3m
Draft:	25.9ft/7.9m
Height:	179ft/54.5m
Passengers:	2,014
Number of Decks:	12 (Including Deck A)
Staterooms:	1,007

SOME FUN FACTS

- 86 per cent of her cabins are outside (have a view).

- 71 per cent of her cabins have balconies.

- Her keel was laid in May 2006.

- She was floated out in January 2007.

- She entered service in December 2007.

- She cost €390 million.

- She was christened by Camilla, Duchess of Cornwall.

- Her maiden voyage was booked out before the ship was completed.

- When she entered service, she was the second largest Cunarder of all time!

GLOSSARY OF NAUTICAL (AND *QUEEN VICTORIA*) TERMS

ABEAM	Off the side of the ship, at a 90-degree angle to its length.
AFT	Near or towards the back of the ship.
AMIDSHIPS (MIDSHIPS)	Towards the middle of the ship.
AZIMUTH POD	A propeller pod that can be rotated in any horizontal direction.
BLUE RIBAND	Award presented for the fastest North Atlantic crossing.
BOW	The forward-most part of a ship.
BOW THRUSTERS	Propeller tubes that run through the width of the ship (at the bow) to help manoeuvrability.
BRIDGE	Navigational command centre of the ship.
COLOURS	The national flag or emblem flown by the ship.
CRUISE SHIP	A ship that carries passengers on itinerary-based leisure voyages.
CUNARDIA	An on-board museum dedicated to the history of the Cunard Line.
DRAFT	Depth of water measured from the surface of the water to the ship's keel.
FORWARD	Near or towards the front of the ship.
HOVE TO	When the ship is at open sea and not moving.
HULL	The body of the vessel that stretches from the keel to the superstructure (*Queen Victoria*'s is painted black).
KEEL	The lowest point of a vessel.
KNOT	One nautical mile per hour (1 nautical mile = 1,852 metres or 1.15 statute miles).

LEEWARD	The direction away from the wind.
OCEAN LINER	A ship that undertakes a scheduled ocean service from point A to point B.
PITCH	The alternate rise and fall of the ship which may be evident when at sea.
PODS	Like giant outboard motors – the Pods hang under *Queen Victoria* and provide propulsion replacing the traditional propeller shafts.
PORT	The left side of the ship when facing forward.
STARBOARD	The right side of the ship when facing forward.
STERN	The rearmost part of a vessel.
SUPERSTRUCTURE	The body of the ship above the main deck or hull (*Queen Victoria*'s is painted white).
TENDER	A small vessel (sometimes a lifeboat) used to transport passengers from ship to shore.
WAKE	The trail of disturbed water left behind the ship when it is moving.
WINDWARD	Direction the wind is blowing.

BIBLIOGRAPHY

BOOKS

Braynard, F.O. and Miller, W.H. (1991), *Picture History of the Cunard Line*, Dover, United Kingdom

Cunard Line (2010), On-board Promotional Material (Various Versions)

Cunard Line (2010), *Queen Victoria*: Technical and Bridge Facts (Various Versions)

Grant, R.G. (2007), *Flight: The Complete History*, Dorling Kindersley Limited, United Kingdom

Miller, W.H. (2001), *Picture History of British Ocean Liners: 1900 to the Present*, Dover, United Kingdom

Miller, W.H. (1995), *Pictorial Encyclopaedia of Ocean Liners: 1860–1994*, Dover,. United Kingdom

PERSONAL CONVERSATIONS

Captain Christopher Rynd

Chief Engineer Martyn Elliot

Cunard Line PR Executive Michael Gallagher

Entertainment Director Alastair Greener

Executive Chef Nicholas Oldroyd

Social Hostess Jennifer Schaper

Third Officer Jack Martin

WEBSITES

Chris' Cunard Page: http://www.chriscunard.com

Cunard's Official UK Homepage: http://www.cunard.co.uk

Cunard's Official US Homepage: http://www.cunard.com

(Courtesy of Alex Lucas)

If you are interested in purchasing other books published by The History Press
or in case you have difficulty finding any History Press books in your local bookshop,
you can also place orders directly through our website

www.thehistorypress.co.uk